# NEW TECHNOLOGY
# sports technology

## Stewart Ross

A+

**Smart Apple Media**

This book has been published in cooperation with Evans Publishing Group.

© Evans Brothers Limited 2010
This edition published under license from Evans Brothers Limited.

Published in the United States by Smart Apple Media, PO Box 3263, Mankato, Minnesota, 56002

Printed by New Era Printing Co. Ltd, China

Library of Congress Cataloging-in-Publication Data

Ross, Stewart.
  Sports technology / Stewart Ross.
    p. cm. -- (New technology)
  Includes index.
  Summary: "Describes the technological advances in the sports industry, including the technology used to create better equipment, sports wear, judging tools, and playing surfaces"--Provided by publisher.
  ISBN 978-1-59920-534-2 (library binding)
  1. Sports--Technological innovations.
2. Performance technology. 3. Sports sciences. I. Title.
  GV745.R68 2012
  688.76--dc22
                              2010044241

June 2011
CAG 1652


9 8 7 6 5 4 3 2 1

**Credits**
Series editor: Paul Humphrey
Editors: Kathryn Walker and Helen Dwyer
Designer: sprout.uk.com
Production: Jenny Mulvanny
Picture researchers: Kathryn Walker
    & Rachel Tisdale


**Acknowledgements**
Cover and title page Guang Niu/Getty Images; p.6 Kerim Okten/epa/Corbis; p.7 Fernando Medina/NBAE/Getty Images; p.8 photogolfer/Shutterstock; p.10 Adidas; p.11 International Tennis Federation; p.12 HEAD UK; p.13 Bill Florence/Shutterstock; p.14 Caryn Levy/PGA Tour/Getty Images; p.15 Herbert Kratky/Shutterstock; p.16 Reuters/Seiko/Corbis; p.17 Fabrice Coffrini/AFP/Getty Images; p.18 Cameron Spencer/Getty Images; p.19 Al Messerschmidt/?Getty Images; p.20 Bjorn Larsson Rosvall/?AFP/Getty Images; p. 21 Hawk-Eye Innovations; p.23 Michael Steele/Getty Images; p.25 Scott Cunningham/Getty Images; p.27 Eye of Science/Science Photo Library; p.29 Adidas; p.30 Marek Slusarczyk/Shutterstock; p.31 Andrzej Burak/Shutterstock; p.32 Karim Sahib/?AFP/Getty Images; p.34 Janek Skarzynski/?AFP/Getty Images; p.35 Liu Jin/AFP/Getty Images; p.36 Philippe Psaila/Science Photo Library; p.37 Sean Aidan/Eye Ubiquitous/?Corbis; p.38 Odd Andersen/AFP/Getty Images; p.39 Jeff Vinnick/Getty Images; p.40 John Zich/AFP/Getty Images; p.41 Drazen Vukelic/Shutterstock; p.42 Sportphotographer.eu/Shutterstock; p.43 bluecrayola/Shutterstock.

# contents

# introduction

During the 2009 World Swimming Championships, world records were broken every day. They tumbled so fast that even the journalists covering the event were not able to keep track. The reason? The sudden speeding up did not come from training or technique but from technology: nearly all the record-breakers were wearing a new type of polyurethane swimsuit.

**Two revolutions** There have been two revolutions in world sports. The first began a little over a century ago and saw the emergence of professionalism supported by vast crowds packed into huge stadiums. The second revolution is going on now. It is happening because technology is being applied to just about every aspect of sports, from the composition of tennis rackets to electronic decision-making to performance-enhancing drugs and therapies.

Some welcome the change, and others hate it. Before judging too quickly, it is worth remembering the enormous benefits that technology brings.

*Technology gone too far? Polyurethane swimsuits in action at the 2009 World Championships, where dozens of records were broken. The suits were later banned because they enhanced the wearer's performance artificially.*

### HOW IT WORKS

Swimsuits made of polyurethane, a form of plastic, contain no textiles. They fit over the body so tightly that they can take half an hour to get on. Once in place, they compress the muscles. This prevents any unwanted muscle movement slowing a swimmer down. Some critics say the suits also trap air inside, giving the swimmer greater buoyancy.

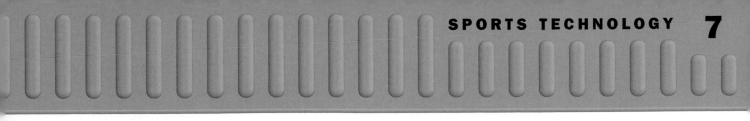

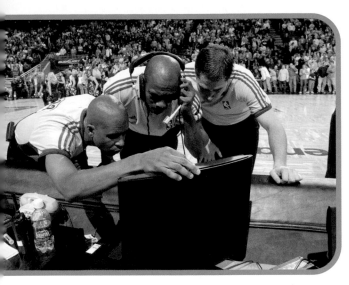

*Referees turn to the video replay screen to check the crucial final shot of a basketball game in which the Houston Rockets overcame the Orlando Magic, 96–94.*

These benefits include a range of new sports, such as skateboarding and hang gliding, greater enjoyment through TV replays, more accurate decisions using video technology, and sporting opportunities for disabled athletes.

**Success to the wealthy** Set against these plusses are the negatives of technology. The most obvious is doping —taking substances that improve performance illegally. There is also the cost of high-tech apparatus such as computerized training programs and machines that measure how efficiently an athlete's body uses oxygen. Such things are so expensive that top-level success is often reserved for wealthy schools, colleges, clubs, and countries.

## FOR AND AGAINST

The pros and cons of electronic decision-making using video technologies.

### For
- Electronic decision-making is more accurate than human referees or umpires.
- Decisions made by machines can be re-run for confirmation.
- Machines are more consistent than people: they do not have "good" and "bad" days.
- Machines cannot be biased in favor of one team or player.

### Against
- Some feel that judging by technology reduces athletics to the level of a computer game.
- Technology cannot yet take into account changing circumstances, such as the altered behavior of a field after rain.
- Referring a decision to technology slows a game down.
- Where technology is used for some decisions but not others, the authority of the human judge is undermined.
- Technology cannot be used in moral decisions, such as bad language or unsportsmanlike conduct.

## CHAPTER 1
# bats and balls

The most obvious impact of sports technology is on the equipment we use. This generally means bats, clubs, sticks, or rackets and the balls they strike. In most sports the effect of high-tech equipment is dramatic. Deep-grooved golf clubs, swinging soccer balls, carbon-fiber tennis rackets, and springy hockey sticks all make modern sports, especially at the top level, faster and more power-based than ever.

**Golf balls** The way a golf ball moves through the air is affected by its composition, weight, size, shape, and surface. To control the impact of technology, therefore, its design is tightly regulated. Engineers could, for instance, design a ball that flies great distances. To prevent this, the sport's ruling bodies have determined how far a golf ball can travel when hit with a specific force. This is the Overall Distance Standard (ODS), which

*Today's golf stars use a type of lightweight carbon-fiber driver. Its power and accuracy threatens to make most traditional golf courses too easy.*

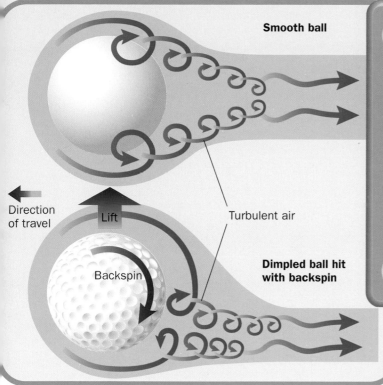

Smooth ball

Direction of travel

Lift

Backspin

Turbulent air

Dimpled ball hit with backspin

## HOW IT WORKS

Contrary to what one might expect, a dimpled golf ball flies farther than a smooth one. This is because when hit with backspin (spinning backwards towards the striker), the air moves more quickly over the upper surface than the lower one. The ball rises to fill the space above it created by the lack of air, traveling faster and farther.

*The dimples on a golf ball scoop up the air and move it to the rear of the ball, where the air pressure remains high so there is less drag holding back the ball.*

applies to all balls. It now stands at 320 yards (290 meters).

The way a golf ball spins affects how it performs. To gain distance, a golf ball needs to spin back towards the direction it has come from. However, this backspin makes it stop more quickly when it lands, which is not helpful for a long shot intended to roll as far as possible. Nevertheless, backspin is essential when trying to stop the ball when it lands near the hole. Remarkably, designers have developed a ball that backspins slowly when hit for distance, but rapidly when the player wants it to stop dead upon landing.

## WHAT'S NEXT?

Scientists are working on golf balls with microchips inside. Using a phone with a satellite link, players would then be able to find a lost ball. The same technology could be used to line up a putt, showing the player the right line for hitting the ball towards the hole when the ground is uneven. This technology, if permitted, would help players line up a shot, but they would still have to hit straight and at the right pace.

**Soccer balls and tennis balls** The modern soccer ball has fourteen panels compared with the old style of 32 panels, and its surface of synthetic materials is virtually waterproof. The outer coating maximizes the friction between shoe and ball. This gives a kicker's shoe greater grip on the ball, allowing them to spin it like a top. The result is a guided missile that is both light and capable of extraordinary flight paths. At the top level, any free kick awarded within 33 yards (30 meters) of the goal is now a potential score.

A tennis ball is a rubber sphere with a felt cover, and its technology is tightly regulated. The tennis balls used for official tournaments have to undergo a series of five tests to make sure they meet approved standards. Because racket power and playing surfaces have changed so much, the modern game uses four kinds of balls. According to the conditions, these tennis balls vary in size, bounce, and the way they change shape under the pressure of a shot. As a consequence, technology and today's tennis are inseparable.

*The match ball for the 2010 Soccer World Cup. For the first time, the outer panels have been molded as parts of a sphere so the ball is perfectly round.*

## BRAIN SAVERS

In the past, a leather soccer ball could double its weight as it absorbed water during a match played in wet weather. This meant heading the ball could give a player brain damage. Modern soccer balls are made of synthetic material or waterproofed leather and gain almost no weight in wet weather. This helps to keep the game quicker and more skillful—and saves the players' brains!

*How a tennis ball is made: 1) A tennis ball starts as a molded rubber half sphere. 2) Compressed air is injected into the center of two halves, and they are sealed together to make a sphere. The sphere is coated in a rubber solution. 3) A covering of mainly cotton backed with a rubber solution is placed over the ball. Then the ball is heated so that the rubber on the ball and the cloth sticks together. 4) The ball is checked for quality and stamped "approved."*

**Sophisticated sticks** Today's field and ice hockey sticks offer power and precision to a degree that was unimaginable even 20 years ago. Much of this is due to the replacement of traditional wooden manufacture with a wide range of synthetic materials including fiberglass, aramids, and carbon fiber. The new sticks produce games that are both faster and harder. In addition, roller skate technology has generated an entirely new form of the game: roller hockey.

**What a racket** The latest tennis racket frames are put together using secret formulas from substances such as ceramics, aramids, boron, and graphite. Today's frames are 300 percent stiffer than older rackets. Because a stiff racket distorts less when striking the ball, the player has more control over where the ball goes. The racket head is also 40 percent larger and the whole racket is 30 percent lighter than previous versions.

## HOW IT WORKS

Some substances, including bone, certain crystals, and a number of synthetic ceramics, are piezoelectric. This means they generate electricity when under stress, and they become more rigid when a current is passed through them. The piezoelectric current created in a tennis racket by the stress of hitting the ball is amplified in the handle and fed back to the frame to stiffen it.

Racket strings are changing, too. Some players say the old strings made of cow gut cannot be improved on. Others have moved over to synthetic equivalents. Even more interesting are the latest anti-vibration measures using piezoelectric crystals. The electric current these crystals generate when the ball is struck is fed back into the ceramics of the frame. This causes it to stiffen, cutting the annoying vibration that may lead to injuries, such as tennis elbow, by up to 50 percent.

**Beefy bats** The size, shape, and composition of baseball bats in professional games are very strictly defined. Only traditional solid-wood bats are allowed. Usually, they are made of maple or white ash. Bat manufacturers must have their bats approved by the

*This series of pictures shows what happens after a ball hits a tennis racket with piezoelectric crystals. The stress of the impact generates an electric current that stiffens the frame.*

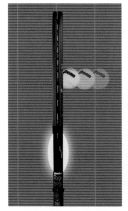 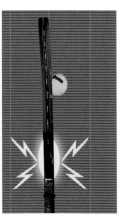 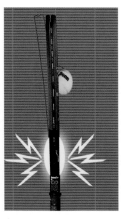 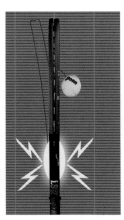 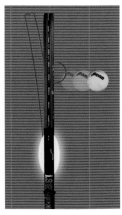

MLB rules committee before a player can use them in a league or exhibition game.

Amateur baseball and softball players have more choices, however, and can take advantage of the technology available to get better hits. These players can choose bats made from all kinds of other things, such as bamboo, aluminum, alloys, and a range of modern high-tech materials such as carbon fiber and aramids. Bats can also be "loaded" or have a weight placed in a certain spot of the barrel to send the ball farther for power hits. Even though there are more choices, each league has rules about the type of bat allowed during games.

## FOR AND AGAINST

Could new technology make baseball a better game?

### For

- Baseball would be a more spectacular sport if players were allowed to use longer-hitting high-tech composite bats.
- A pitcher could throw more unplayable curveballs with a shinier baseball.

### Against

- Baseball is about a player's skill, not how good their equipment is.
- Once new technology is allowed, one cannot predict how the game will change.
- Different bats and balls would make all existing statistics and records irrelevant.

*Of all the major sports, baseball has probably most fiercely resisted the impact of technology on its equipment.*

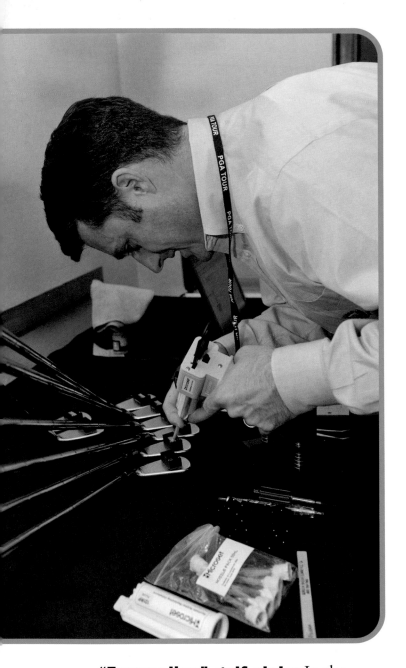

*A scientific researcher checks the grooves on golf clubs to make sure they conform to the official standard. Tiny variations in the grooves' size, shape, and spacing can make all the difference in a tricky shot.*

**"Trampoline" golf clubs** In the professional game, the shafts of golf clubs are made from carbon fiber reinforced polymer, giving the best strength-weight ratio. The heads are of three kinds: drivers for sending the ball a long way, irons for lofting the ball so it drops close to the hole, and putters for rolling it into the hole.

Technology could easily produce a driver capable of sending the ball more than 500 yards (450 meters). This would make every fairway in the world unnecessary because players could hit the ball to the green with one shot rather than the two or three needed today for most holes. Modern drivers are hollow with a face that acts like a springy trampoline when striking the ball. The spring of a club head is measured as its CoR (Coefficient of Restitution). To prevent golfing stars from sending the ball right up to the hole with every drive, a club head's CoR may not be more than a set figure (currently 0.83 CoR).

Irons have grooves along the club face to add the grip needed to get the ball spinning. By 2009, scientists had worked out the optimum spacing and shape of grooves to give players the greatest advantage when striking the ball in difficult situations, such as from long grass. This changed the art of golf so much that steps were taken to ban clubs with U-shaped grooves, known as U-profile grooves.

**Track and field equipment** There are two areas in which technology has revolutionized track and field events: the javelin and the pole vault. The modern hollow javelin could be thrown beyond the central grassy area if technologists had not made it more difficult to "fly." They did this by moving its center of gravity (the point of balance) away from the middle of the shaft. This made the javelin less aerodynamic, meaning its weight distribution and shape reduced the time it was in the air.

By allowing the pole to be made of any substance, the pole vault regulations are an open invitation to engineers to come up with something special. And they have. Today's pole is

*The remarkable heights scaled by modern pole vaulters would be impossible without flexible fiberglass poles and artificial run-up surfaces.*

## HOW IT WORKS

Air flowing over a javelin gives it a slight lift. This is greatest at the "center of pressure," or the place where the upward pressure is strongest. Because the center of gravity is now in front of the center of pressure, gravity pulls the nose of the javelin down. This reduces the length of a throw and makes a nose-first landing more likely, which is easier to mark.

a specially shaped and super-bendy spring of fiberglass or similar material. The result? Records that stood at about 3.25 m (10.7 ft) for men have today soared to more than 6 m (19.7 ft). Women's heights have gone up from about 4 m (13 ft) to 5 m (16.4 ft).

## CHAPTER 2
# judging and timing

In 2004, a South African soccer referee shot dead a coach who had questioned his decision. It may be extreme behavior, but not that surprising. In major sports, passions run very high and officials are under extreme pressure. Many engineers believe that technology could make the situation easier because most sports judgements are better made by machine than by fallible humans. Some say we should go even further and make the human referee unnecessary.

**First past the post** A modern photo finish uses a high-speed digital camera taking 2,000 images a second. An object as small as a human hair can be detected. Even if a human eye could get near such precision, the brain behind it has no playback or freeze-frame facility for double checking. Cameras win every time.

**False starts** Technology improves track starts, too. The starting blocks from which sprinters push off are wired to the starting gun. If the block senses

*Photo records of the finishing line, like this one showing Usain Bolt breaking the world 100 meter record in 2009, serve two functions. They separate the athletes as they cross the line and provide precise split-second timing.*

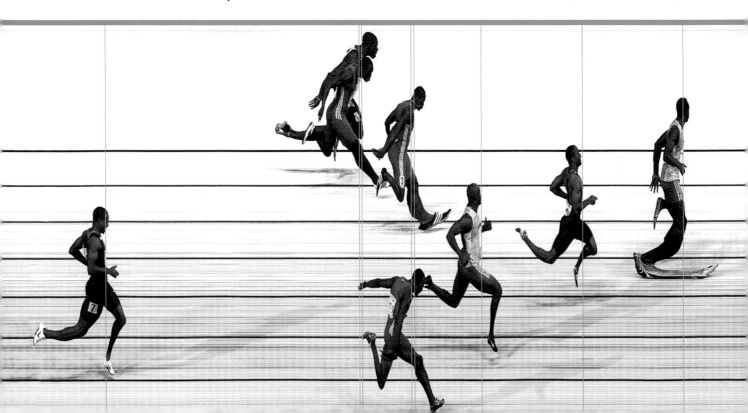

## HOW IT WORKS

At sea level and in dry air at 68 °F (20 °C) sound travels at 768 miles per hour (1,236 km per hour). A sprinter on the outside of a race track, therefore, hears the starter's gun a fraction of a second later than a competitor on an inside lane. To ensure that all runners hear the gun at precisely the same time, each starting block has a speaker that relays the starter's commands.

*A modern starting gun is wired to timing devices and to speakers behind the starting blocks in each lane.*

pressure in less than 0.11 seconds (the fastest a human being can react) after the gun sounds, this means the athlete has started early. When this happens, a false start is recorded electronically against the offending lane.

Technology scores also when it comes to distance measurement. A reflector placed in the ground where a shot, discus, or javelin lands bounces back a laser beam, and the throw's precise length is flashed up in a split second.

**Chips to the rescue!** A stopwatch might be fine for timing your personal training, but it's not much use when it comes to world records. Clock

technology is also hard pushed to cope with staggered starts, as in the Tour de France cycle race, or with thousands of runners pouring over the finishing line in the Boston or New York Marathons.

Top-class races in everything from running to skating, cycling to motor-sports, are timed by devices linked to the photo-finish mechanism, capable of giving times to the nearest 1/100 of a second. Staggered or mass start events,

such as a triathlon, are timed by attaching a chip to part of each competitor's clothing. For example, timing chips are often attached to competitors' shoelaces.

**Timing swimmers** Because water often obscures sight lines, swimming has developed a different technology. Sensors in the blocks record the start time. To complete a race and record a time, a finisher must touch a plastic pad. This instantly records the lane and the time. In 1987, the first time this equipment was used, the swimmers could not believe the times and turned violently on the judges. But a quick check showed that the equipment had worked perfectly.

*What the eye doesn't see, the touch pad will record. In the pool, times and positions are recorded when the swimmer touches an electronic pad at the end of the race.*

## TRIPPED CHIP

Finishing first in the 2006 Chicago Marathon, Kenyan runner Robert Cheruiyot tripped on a mat placed just before the finish, fell forward, and knocked himself out. His body was over the line, but his shoelaces with a timing chip embedded were not. Cheruiyot was not declared the winner until his time was confirmed by an old-fashioned stopwatch.

**See it again** The video replay is a key element of the modern sports scene. It has a double function. The more fast-paced sports, such as hockey, use it as a way of checking decisions, while in others, it is just part of the TV coverage. Replay evidence can also be used to show foul play and cheating. Sometimes, a video screen can help competitors. Glancing up at the screen while running a race, for example, lets runners see where the others are.

In sports where video replays are used, the pause, followed by the moment when an announcement such as "goal" or "foul" flashes on a giant screen above the crowd, creates excitement by adding an element of suspense. The decision is also correct, reducing bitter post-game arguments. This is why video replay is permitted on a limited basis in several sports, including football, basketball, and tennis. Some sports—most notably soccer and baseball—are reluctant to adopt decision-making technology. They feel it would slow the game down and undermine the referee.

**Appeals** Replay technology has introduced another more controversial element to sport. This is the appeals system used in football, basketball, and tennis. The idea is that a team (usually by the coach) or an individual player is given a limited number of appeals against decisions that have not gone their way.

A good example is when a tennis player challenges a line judge's call. The appeal is referred to an official with access to technology. Although the decision is technically correct, the system introduces the tactic of appealing to destroy the concentration of an opponent.

*The NFL allows limited appeals to a digital replay facility. Here, a referee checks an incident on the video replay before making a decision.*

## FOR AND AGAINST

Does appealing to technology improve or ruin sports?

### For

- Allowing players to appeal against decisions adds a new and exciting dimension to sports.
- Appeal to technology keeps human referees and umpires on their toes.
- The possibility of an appeal to technology makes it more likely that game-deciding decisions will be correct.

### Against

- Appeals encourage bad sportsmanship as they can be used to unsettle an opponent.
- The authority of officials is undermined by an appeals system.
- Technological appeal is neither one thing nor another: if technology is always right, then use it all the time; if it is not, then don't use it at all.

*Many argue that allowing tennis players to challenge the officials' calls, as Argentina's David Nalbandian is doing here, undermines their authority and makes the sport look bad.*

**Subtler and subtler** Digital video evidence is old news. The latest decision-making machinery is based on military technology used in missile tracking systems. Hawk-Eye, the best-known version, is used in professional tennis. It employs a battery of at least six smart cameras to track and record the flight of a ball. Similar technology

## HOW IT WORKS

Hawk-Eye works by feeding a number of consecutive images into a highly sophisticated computer program that combines them into a single 3D moving sequence. This is then relayed to a screen. The program shows precisely what has happened, such as exactly where a tennis ball has landed.

is used to record every movement made by a player during a match, allowing individual performances to be analyzed in great detail.

We now have the digital technology to do everything an official does, and better—apart from punishing foul language and unsportsmanlike behavior. Perhaps it won't be long before we have a robot capable of dealing with those issues, too.

*The precise path of a tennis ball can be recreated on screen using digital cameras and computers. In the future, tennis may allow all important decisions to be made objectively by machine. This could prevent arguments, and referees and line judges may no longer be needed.*

## WHAT'S NEXT?

On a baseball park or soccer field stand robots powered by solar panels. The decisions they give are correct every time and the players appreciate this. In 50 years' time, robot umpires may well be the norm in professional events. Would we lose anything? Some believe we would lose the excitement of controversy, as well as the personal interaction between player and official.

ROLEX OFFICIAL REVIEW

# CHAPTER 3
# surfaces and stadiums

Technology in the arenas gives the modern sport near-perfect conditions. Football fields are soft like carpet, and running tracks are designed for record-breaking. Even swimming pools have technology that benefit the athletes. Spectators, too, benefit from comfortable seats, TV screens, and uninterrupted views of the action. Meanwhile, thoseat home can watch the entire event in high definition from their living rooms.

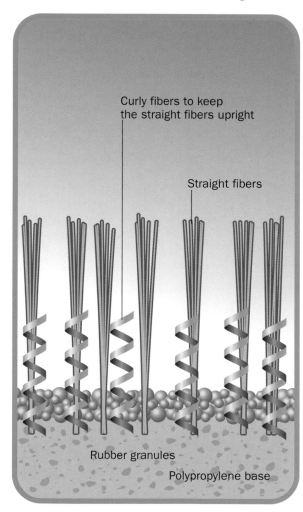

Curly fibers to keep the straight fibers upright

Straight fibers

Rubber granules

Polypropylene base

*Many believe that it is only a matter of time before all professional sports, especially at the top level, are played on artificial surfaces like this.*

**Grass: real and plastic** If there is one area in which technology has without a doubt improved sports, it is playing surfaces. Today, there are fields that remain green and firm year-round.

Smooth playing surfaces are achieved either by using completely artificial grass or by mixing grass with plastic

## HOW IT WORKS

Many modern sports fields are a mixture of grass and plastic. This is so subtly done that the grass-plastic mix is invisible to the naked eye. To achieve this, the field of natural grass is sown by a computer-controlled machine that inserts millions of silicon-lubricated polypropylene blades among the roots of the natural grass. The result is the best of both worlds: the feel and cost-savings of grass with the durability and drainage of an artificial field.

fibers on a heated, well-drained, more traditional field. When tall stadiums block out sunlight, ultraviolet lamps are used to keep the grass green and healthy. Modern drainage systems mean that downpours no longer make surfaces unplayable.

**Hard surfaces** Two surfaces that owe everything to technology are found on synthetic running tracks and artificial tennis courts. Running surfaces vary from the older Tartan tracks of pure polyurethane to the latest mix of rubber granules in a polyurethane binder on a firm base. These tracks drain rapidly after rain. They are extremely expensive, however, widening the sports gap between the rich nations and the rest of the world.

## WHAT'S NEXT?

In the 1970s, many football and baseball stadiums switched from natural grass to an artificial grass called AstroTurf. Since 2000, fields are being converted to a newer, softer artificial turf called FieldTurf or going back to the natural grass. Players tend to favor natural fields, but the FieldTurf provides easier upkeep, and makes conversions to stage shows or other sporting events less challenging.

*The surfaces of top synthetic running tracks are kinder on runners who trip and fall, reducing the risk of serious injury.*

## FOR AND AGAINST

In recent years, all top-class field hockey has been played on artificial fields regularly soaked in water. These water-based fields have caused much controversy.

### For
- Hockey players say that a water-based field offers a more even and faster playing surface.
- Clubs and national associations have spent fortunes installing water-based fields, and they don't want to change them yet again.
- Top players have adjusted their game to the water-based field.

### Against
- Water-based fields are extremely expensive to install and maintain.
- Environmentalists say precious water should not be wasted on playing fields.
- Countries like India and Pakistan, where water is not plentiful, are at a disadvantage if water-based fields are the norm.

The modern range of high-tech tennis court surfaces makes the game more varied and more interesting. Alongside traditional grass, still used in Britain's Wimbledon tournament, there are hard, clay, and wooden courts. Hard courts have a concrete or asphalt base topped with acrylic or similar substance. Clay courts actually have little or no clay in them and have not for many years. Instead, the base is topped with some sort of rubberized carpet layer. Wooden surfaces are most likely to be found at indoor courts where weatherproofing is not an issue.

**Splash and dash** Scientists involved in swimming pool designs have come up with all kinds of devices to cut inter-lane disturbance. These include lane dividers and pool edges that absorb waves, and recycling pumps that maintain a smooth surface on the pool. All these improvements do not make much difference to the average person going to their local

## ARTIFICIAL ATHLETE

The harder a running track, the faster athletes can go. To prevent things from getting out of hand, a track's thickness and bounce is measured by an "artificial athlete." This is a piece of machinery based around a metal tube with a spring inside. The running surface has to meet the standards set by the IAAF (International Association of Athletics Federations) in order for performances on it to be recognized.

phase change materials (PCMs). These are gels or waxes that absorb heat without gaining much heat themselves. Minute capsules of PCMs are incorporated within the fabric of sportswear at the manufacturing stage, giving the wearer a much greater chance of staying cool in competition.

**Full support** Elastane, a synthetic fiber, is present at just about every sports event. Because every thread is elastic, it combines support with flexibility. By holding muscles steady, it also helps blood flow and calms vibration.

*A close-up picture of the surface of a shark-skin swimsuit. The minute ridges mimic those on a shark's skin, reducing turbulence and therefore reducing drag through the water.*

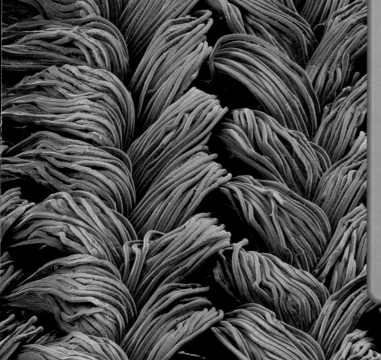

Similar claims are made for the shark-skin-type materials used for swimsuits. These are based on fish skin that has millions of minute, tooth-shaped scales all lying in the same direction.

## FOR AND AGAINST

In recent years, high-tech swimsuits have caused enormous controversy in pools around the world.

**For**

- Shark-skin and polyurethane swimsuits improve performance, allowing records to be broken. This increases the sport's popular appeal.
- The new swimsuits are more up-to-date, making the sport more attractive to TV audiences.
- A sport that turns its back on technological development presents an old-fashioned and out-of-date image.

**Against**

- Shark-skin-style swimsuits alter the characteristics of the human body, changing the nature of the sport.
- High-tech swimsuits are very expensive, which benefits athletes from wealthy countries.
- If winning or losing is a matter of technology, fair play and good sportsmanship are undermined.

They create a minimum of turbulence when water passes across them in the same direction as they lie.

When it comes to toughness, modern synthetic fibers take all the beating they can get. Combinations of cotton, nylon, elastane, and polyester are almost indestructible, so that ripped clothing—even in the toughest contact sports—is just about a thing of the past. There are football shirts designed to stretch huge distances when pulled, making it easier for referees to spot foul play. Even more revolutionary, "sheer thickening" materials like d3o toughen up only when they are required to do so.

Many activities—especially winter sports, car racing, and football—require strong physical protection, and the

*Sheer thickening materials consist of molecules that can change their properties in different situation.*

## HOW IT WORKS

When honey is stirred, it becomes harder the quicker the spoon is moved. Sheer thickening materials work in the same way: the greater the force exerted upon them, the harder they become. The change happens in 1/100th of a second. Padding is soft and pliable until hit. Then it becomes rock solid. These materials are perfect for contact sports like football and rugby, and very useful for the helmets worn by cyclists and horse riders.

proper sportswear can be hugely beneficial. Stock car racing is an extreme example. A many-layered overall suit enables drivers to survive temperatures of 1,560° F (850° C) for more than 30 seconds yet still remain flexible enough

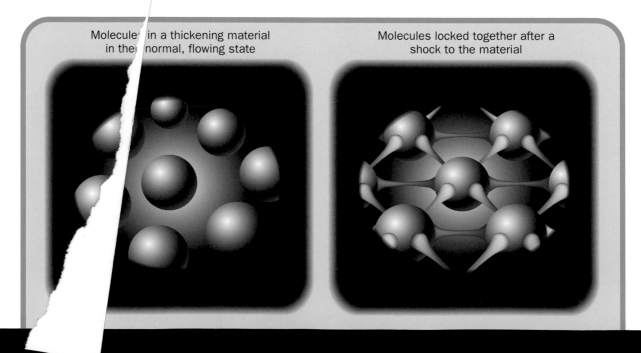

Molecules in a thickening material in their normal, flowing state

Molecules locked together after a shock to the material

to drive at 200 mph (320 kph)! Helmets are crucial to prevent injury from crash impact or flying debris. The outer layer is usually a shell coated in a resin made from carbon, glass, and an aramid used in bullet-proof vests. To further absorb impact, a layer of polystyrene or polypropolene foam wraps around the head. Finally, an inner flame-resistant lining protects the driver in case of a fiery crash.

## PERFORMANCE RATING

Sports scientists are working on athletic shoes that contain movement sensors and an electronic chip. The chip is linked to a computer when the training session is finished and the athlete can read off detailed post-training printouts of their performance: how far they have traveled, how fast, stride pattern, etc. The information can be linked wirelessly to a computer and become part of a comprehensive training program.

*The extremely pricey intelligent sports shoe has a built-in microchip that calculates the wearer's weight, pace, and terrain and adjusts the cushioning accordingly.*

**Feet first** The biggest impact of sports science on footwear has been to produce items designed specifically for each sport. All top level athletes wear highly individual shoes.

Despite all the hours of research, there is still disagreement over what design works best. Soccer shoes, for example, have been discussed frequently regarding whether studs or blades should be used on the bottoms. Each has its supporters, and researchers have yet to decide which gives the best possible combination of grip and quick release to avoid injury.

# CHAPTER 5
# sports machinery

All sports involve technology at some level, even if it's just a starting line and a finishing post. But many of today's most exciting and popular sports—car racing and skateboarding, for example are actually created by technology. Technology has also provided new ways of playing old sports, such as roller hockey and waterskiing, and has opened an entirely new world of sport for those with disabilities.

**New horizons** Every weekend, millions of people take part in sporting activities—on land, water, or even up in the air. Kids rumble through terrain on skateboards, BMX riders hurl themselves around muddy tracks, and on the lake, water-skiers jump and dance behind speedboats. Above, freefall parachutists float from a sky alive with hang gliders, paragliders, and buzzing microlights. In short, technology + sports = something for everyone.

*David Weir leading the London wheelchair marathon. No athletes have benefitted more from modern technology than those who were previously excluded by their disabilities.*

## WHAT'S NEXT?

Rowing boats with more than one body (known as a "hull") would be more stable and therefore, probably quicker. It would also be more efficient to have the pivots that hold the oars moving backward and forward rather than the rower doing so on a sliding seat. Technology could bring both these improvements. It is only regulation that prevents change.

## HOW IT WORKS

A paraglider is basically a huge fabric wing with a person suspended beneath. The wing is made of two layers of fabric joined to form cells like a honeycomb. The front edge of most cells is open, the rear closed. The wing holds its shape because air pushes into the cells and keeps them inflated like a balloon.

### Super machines on four wheels

Sports technology reaches its pinnacle in Formula 1 racing (F1). F1 cars use only a 2.4 liter engine, the same size as a family sedan. Amazingly, by reaching the maximum permitted 18,000 rpm (about three times the revs of a normal street car), the F1 racing machine achieves a stomach-churning 220 mph (360 kph) down the straightaway. On a tight bend, the drivers' bodies may experience a force equivalent to over five times the force of gravity (5Gs), close to the maximum a body can sustain.

In the interest of cost, safety, and fairness, F1 strictly controls the materials used in the cars, their aerodynamic design, and electronic input. This has been an area of much controversy over the years.

*The joy of paragliding. Technology does not just improve standards in traditional sports, it creates new sports and widens popular choices and participation.*

## WHAT'S NEXT?

The noisy, polluting internal combustion engine is doomed. It won't be long before all cars are electric, and that will include the racing versions. But could you have a battery-powered race car that reaches speeds of 200 mph (320 kph)? It's certainly possible. The speed record for an electric vehicle is already around 220 mph (350 kph) over a short distance. Technology will soon enable such speeds to be maintained throughout an entire race.

One of the most interesting developments pioneered by F1 was the use of computers to get the maximum performance from a car. Before the system was outlawed in 2007, computerized traction control applied the perfect grip of the tires on the road, replacing the driver as the judge of precisely how much power to apply at any given moment. This gave the best possible traction by eliminating inefficient wheel spin and the possibility of going out of control.

*Technology is so essential to the success or failure of a grand prix car that critics say car racing at this level is not really a sport at all.*

## Super machines on two wheels

By comparison with F1, motorcycling and bicycling operate on small budgets. Nevertheless, the technology behind a 4-stroke 800 cc MotoGP bike (the motorcycle equivalent of F1) is right at the cutting edge. These bikes can power up to nearly 220 mph (350 kph).

To keep a machine on the track, tire technology is as important as engine, aerodynamics, and suspension. Both F1 and MotoGP allow a mix of soft (good grip, but quick wearing) and hard (longer-lasting, but less grip) tires. The difference between four- and two-wheel adhesion is shown by the fact that a MotoGP bike circles the Spanish Jerez track in about 1 minute, 40 seconds, while an F1 racer takes around 1 minute, 17 seconds.

Obviously, you won't find those sort of speeds in bicycle races. Nevertheless, the technological revolution has also worked its sporting magic in this field. The advances have been controversial, too, because teams that spend the most money can appear to get the best results. Track racers, road racers, mountain bikes, BMX—there's a machine for every bike sport. Many bikes have frames of lightweight metals, such as titanium, carbon-fiber components, and disc brakes.

### HOW IT WORKS

Normal cycle brakes work by forcing hard rubber pads against the rim of the wheel. Friction causes the bike to slow down. Disc brakes increase the amount of friction and are therefore more efficient, as well as less damaging to wheel rims. They operate with much larger pads gripping on either side of a large metal disc attached to the hub of the wheel. Because they are closer to the wheel hub they work better in wet weather, too.

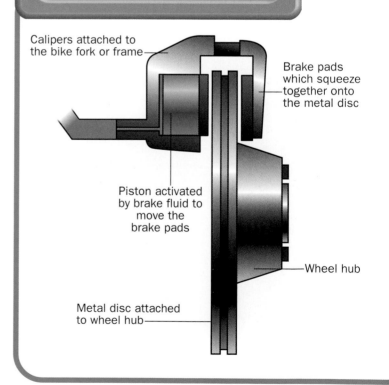

Calipers attached to the bike fork or frame

Brake pads which squeeze together onto the metal disc

Piston activated by brake fluid to move the brake pads

Wheel hub

Metal disc attached to wheel hub

*Disc brakes increase the surface area of friction, which offers greater stopping power to any machine to which they are attached.*

*The modern track cycle is extremely high-tech, but still limited by the regulation that the rider must be above the bike and not in a more streamlined recumbent (lying-back) position.*

The bicycle is the most efficient self-powered means of transportation, but it could be improved. Engineers know that sitting on a bike causes more drag than lying back on it. Also, cyclists in a lying-back, or recumbent, position use their leg muscles more efficiently. The change to recumbent bicycles could happen if the cycle racing authorities give it the green light.

## FOR AND AGAINST

How much should cycling performance depend upon the technology of the machines and the rider's equipment?

**For**
- The technology that enabled Britain to win twice as many cycling medals as any other nation (14) in the 2008 Olympics has advanced the sport.
- Every country is free to develop similar technology.
- Cycling is a sport based on technology, so technical limits are artificial.

**Against**
- Britain's $41 million funding for cycling gave it an unfair advantage over other nations, especially developing countries.
- Sports are about human skill and determination, not who can build the best machine.
- Cycling events would be better if all competitors used identical machines.

**Sports for all** A modern wheelchair, with a frame of lightweight aluminum reinforced with tough titanium, now weighs around 18 lb (8 kg). Just a few years ago the equivalent machine weighed 50 lb (23 kg). Modern versions can be built with a fifth wheel for greater stability and the seat and

## SPRINGY LEGS

South African athlete Oscar Pistorius (born 1986) had both his lower legs amputated at age 11 months. Technology came to his rescue when he was given artificial lower limbs. Two L-shaped carbon fiber springs were strapped on just below the knee. Conflicting studies on whether Pistorius has an advantage over able-bodied runners have prevented him from competing with them. His accomplishments have been limited to races with other amputees.

rests are fully adjustable. These developments have opened up all kinds of opportunities for disabled athletes.

No longer confined to track and field events, the modern disabled athlete can take part in basketball, fencing, tennis, and even soccer and football. For contact sports, the chairs are built with guards and wheels that lean steeply inwards for extra stability.

Technology has also opened up the world of winter sports to disabled athletes, allowing them to speed down the slopes on specially constructed skis.

*Some scientists suggest that the artificial limbs worn by South African runner Oscar Pistorius give him greater spring off the track than an able-bodied athlete.*

There is also the more dangerous fourcross bike, a four-wheeled mountain bike in which disabled athletes careen down mountains and leap across ravines.

# CHAPTER 6
# training and cheating

Training for all serious sports is a tightly organized scientific process that works every aspect of an athlete's body and mind. At the top level, the boundaries of what is and is not permitted are often blurred.

**University level** Around the world, most self-respecting universities boast a sports science department. As a result, every avenue to sports success has been analyzed and plotted, and there are almost no top-class athletes whose lives are not arranged by others.

*Motion capture: the athlete wears white reflective patches that enable his movements to be tracked accurately by video camera. Close analysis of the images will enable coaches to suggest how style and technique can be improved.*

Motion analysis involves analyzing a person's technique on computer with a digital camera. It is now basic to most training. Motion analysis is associated with more complex areas of study, such as biomechanics, which is the study of the way forces act on or through the human body. Kinanthropometry, using data to predict a body's athletic potential, is another new area of research.

Another example of the influence of sports science is the fairly recent

## HOW IT WORKS

Motion analysis takes many forms. At its simplest, it involves taking video of an athlete in action, perhaps from several angles, then replaying the movement in slow motion. This reveals weakness in technique that can be corrected. The athlete's performance may also be fed into a computer program that compares it with a perfect model. This way, the athlete can see ways to improve their technique.

distinction between "fast twitch" and ordinary muscles. Fast twitch muscles are essential for sprinting. In fact, it is now possible to analyze how much of each muscle type an athlete has and, before they run a race or even train, predict how well they will be able to perform.

**Training** Training is no longer about just getting fit. Fitness is measured in precise detail. The most common device is the VO2 max test that measures the rate a body absorbs and uses oxygen.

Hopefuls who train and still perform poorly on the max test might as well give up before they go any further: they were simply not born with a good enough heart-lung system.

A large part of athletic performance is linked to the body's ability to get oxygen from the lungs to the muscles as quickly and efficiently as possible. This task is performed by red blood cells. The body's production of red

## HOW IT WORKS

The VO2 max test requires an athlete to do exercise that gradually becomes more strenuous. This is normally done on an exercise bike or a treadmill. The subject wears a mask with tubes attached through which all the air they inhale and exhale passes. The tubes are linked to machinery that measure the amount of oxygen and carbon dioxide present in the air that is breathed in and out. VO2 max (level of fitness) is reached when the amount of oxygen the body uses remains the same even when the level of exercise increases.

*The VO2 max test measures the efficiency of an athlete's heart-lung system. This reflects both the effectiveness of the system with which they were born (something that can't be changed) and their fitness (which can be improved).*

blood cells is regulated by the hormone erythropoietin, better known as EPO. Basically, the more EPO in the body, the more red blood cells that are produced, which sends more oxygen to the muscles.

EPO can be taken—illegally—as a performance-enhancing substance or it can be increased naturally by living at an altitude over 5,740 feet (1,750 m). At such an altitude the body senses the lack of oxygen in the air and produces more red blood cells. These remain in the system for a time even after the athlete has returned to a normal altitude. This is known as altitude training. It can be mimicked by using an oxygen tent.

## HOW IT WORKS

An oxygen tent should, in fact, be called a low-oxygen tent. It is air-tight and is filled with circulating air from which some of the oxygen has been removed. This mimics the condition at a higher altitude where lowered oxygen levels cause the human body to produce more oxygen-carrying red blood cells. For a while, these cells take more blood to the muscles when normal conditions are restored, which also aids in recovery from injury.

**Diet** "You are what you eat," the saying goes, and it is especially relevant to athletes. Food technologists have worked out that the ideal athletic diet is provides 60 to 70 percent of calories from carbohydrates, 12 percent from proteins, and 18 to 28 percent from fats. Before an event, it is advised to go "carbohydrate loading." This means stocking up on calories from pasta or potatoes.

*Most athletes eat bananas to give them extra energy. The way the body converts food into energy is now so well known that all top athletes follow carefully planned diets.*

## FOR AND AGAINST

The use of performance-enhancing drugs causes more argument than any other issue in sports.

### For

- Some experts argue that all ways of improving performance should be allowed because they produce more skilled and exciting events.
- As it is impossible to detect all drug abuse, it would be more fair to let athletes take what they want.

### Against

- Taking substances that artificially improve performance is cheating.
- All athletic drugs carry serious health risks.
- If drugs were made legal, all current athletic records would become meaningless.

**Drugs** It seems there are more ways of cheating at sports than there are sports themselves. Many involve taking substances that are banned because they artificially improve the body's performance, giving an athlete an unfair advantage. All have seriously damaging side effects.

Testing for drugs is done by analyzing samples of an athlete's urine and blood. The exact testing processes are secret, but we know blood testing, for example, involves separating blood cells from blood plasma by spinning blood in a centrifuge. In some countries, this is done randomly all year; in others it is done

*Sample testing for the 2010 Winter Olympics inside the anti-doping lab at the Olympic Oval, Richmond, Canada. Testers are always struggling to keep up with the secret laboratories that make a fortune from the manufacturing of more and more sophisticated products to enhance performance.*

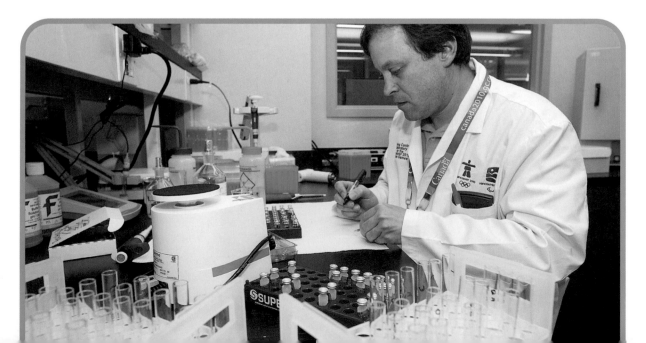

only in the competition season, enabling athletes to load up on illegal substances in training but produce negative test results when competing.

The earliest performance-enhancing drugs included stimulants such as cocaine, as well as amphetamines that increase heart rate. Anabolic steroids that build muscle and power have been around for some time, too, but unscrupulous scientists are always producing new ones that are difficult to detect. The ever-growing list of banned steroids is now two pages long. It includes substances like zilpaterol, officially produced to fatten cattle, and nandrolone, which occurs naturally in the body in small quantities, but is regarded as illegal if more than the normal amount is found in the urine. Since new steroids, legal and illegal, are produced every month or so, there is a continual game of cat and mouse between the drug producers and the testing authorities trying to find ways to detect the new substances.

Technology has also come up with various ways to cheat at a drug test. The cruder ones involve artificial body parts, such as hidden bladders filled with clean urine. Athletes can also take masking agents that hide the presence of illegal drugs in their system.

More sophisticated methods of cheating include blood doping. This is

*This poster is protesting against steroid use by U.S. baseball players. Opinions about what athletes should and should not be allowed to take vary widely. Some argue that adults should be free to take anything they want, and accept the consequences if the long-term effects are harmful.*

when an athlete's blood is taken from their body, concentrated to increase its red blood cell count, and then returned by transfusion to the body. A similar

## WHAT'S NEXT?

Scientists are close to being able to grow human body parts from stem cells (cells capable of developing into all types of cells required in a living organism). From here the next step will be gene therapy. A football player with a worn out knee or a javelin thrower with a badly damaged shoulder will be able to grow a replacement. Where will it end?

process is used with hormone doping. This involves giving athletes hormones, often harvested from dead bodies, that trigger things like growth or red blood cell production.

The ultimate technological cheat is to get the body to produce performance-improving substances on its own by genetic manipulation. If and when it does happen, testing for artificial performance enhancement will become virtually impossible.

## HOW IT WORKS

Genetic manipulation or engineering in athletics means transplanting a human gene (basic unit of heredity in a living organism) into the body that produces a performance-enhancing protein or hormone. The process has worked with mice, enabling them to gain an extra 25 percent of muscle in just three weeks. Officially, however, no one is quite sure whether it has yet been tried on people. Its consequences are potentially terrifying as it could cause changes to the mind as well as to the body.

*Weight-lifters and body-builders (shown here) are the two kinds of athletes most frequently accused of using illegal muscle-building substances such as steroids.*

# conclusion
## the sports industry

Technology has benefitted sports in countless ways. It has raised performance, increased safety, created new sports, and brought sports to those previously excluded. At the same time, it has turned sports into an industry, made top-level participation incredibly expensive, and through the possibility of genetic doping, introduced a new world of serious controversy. Depending on one's point of view, the future is either rosy or dark.

**Plus and minus** The enjoyment that an amazing football player like Peyton Manning brings to sports fans is possible only through technology.

His fitness, the carpet-like field, the state-of-the-art shoes, the live TV coverage—all these things result from the application of science and technology

*Equipment and clothing technologies have helped to raise standards and improve safety in winter sports such as skiing. Technology has also made it possible to create artificial snow at popular ski slopes.*

### AN END TO RECORDS

The French medical research company IRMES figures that top athletes perform at close to 99 percent of the body's capacity. One hundred years ago this figure was 75 percent. As the present rate of progress continues, we will get even closer to 100 percent by 2050. Some believe that by around 2060, there will be no more world records unless genetic engineering is used to produce "super athletes" designed for optimum performance at a particular sport or event.

to sports. Such examples could be multiplied a thousand times over.

Add to this the explosion of disability athletics and the host of new sports, and the case in favor of technology seems unanswerable. But it is not.

In 2008, the top 20 countries on the Beijing Olympics medal table featured only two developing nations, Jamaica and Kenya. The message is clear: because the application of sports technology is so expensive, medals go to the wealthy athletes. Technology only emphasizes the gap in the sports world between the haves and the have-nots.

Sports are essentially about fun: exercise and competition for the sheer joy of it. Technology, in the form of broadcasting, equipment, and facilities, has helped to undermine this attitude by turning sports into a business. Because so much money is involved, winning at the professional level has become the sole purpose. Sadly, this attitude filters down to the playground. Even for children, winning the game can come before sportsmanship.

Finally, there is the grim reality of doping. The list of those involved in doping scandals grows almost daily. And insiders say we don't know half of what really goes on.

*Track athletes from Africa dominate longer track and road races, but lack of money means they cannot succeed in expensive high-tech sports.*

## WHAT'S NEXT?

The market in steroids and other doping agents is worth several billion dollars. The budget of the World Anti-Doping Agency (WADA) is $25 million. Guess who wins? One answer could be to levy an anti-doping tax on all sports transactions. This would give WADA the funds it needs to do its job adequately.

# glossary

**3D** three-dimensional; having length, width, and depth

**acrylic** a synthetic fiber

**aerodynamics** the science of air flow around a physical object

**aramid** an aromatic polyamide; a strong, heat-resistant synthetic fiber

**biomechanics** the science of the forces acting on or through the human body

**boron** an inactive element

**carbon fiber** graphite (carbon) in the form of fiber

**centrifuge** a machine that rapidly spins a liquid in order to separate out its constituent ingredients

**ceramic** an inorganic, non-metallic solid, often containing silicon

**composite** made of several materials

**d3o** lightweight and flexible materials that become hard on impact

**doping** taking substances that artificially enhance performance

**drag** the resistance acting on a body as it passes through a liquid or gas

**drivers** golf clubs for long distance hitting; also known as woods

**elastane** a stretchy synthetic fiber used in clothing

**EPO** the hormone erythropoietin that stimulates red blood cell production

**F1** Formula One, the global car racing organization

**fiberglass** a material made from fine fibers of glass mixed with a plastic

**free kick** a soccer rule that allows a team that has been fouled to take possession of the ball and kick it anywhere they want, even at the goal

**gene** the complex chemical structure that is the basic building block of all forms of life

**graphite** carbon in its most stable form

**hormones** the body's chemical messengers

**irons** angled golf clubs for hitting short distances with precision

**piezoelectric crystals** crystals that produce electricity when stressed

**polymer** a large molecule with a structure in repeating units; DNA is the best-known natural polymer; synthetic polymers are often plastic

**polypropylene** a synthetic plastic polymer

**polyurethane** a synthetic plastic polymer containing urethane

**silicon** a non-metallic element that combines with oxygen and a metal to form one of many useful minerals

**steroids** drugs that mimic the male hormones and help build muscle

**synthetic** artificial

**tennis elbow** pain in the lower elbow usually caused by repetition of a vigorous action such as striking a tennis ball

**ultraviolet** light beyond the violet end of the spectrum, having radiation wavelengths less than that of visible light

**water-based field** a hockey field with an artificial surface that needs to be heavily watered

# **more** information

## Books

*Sports and Sporting Equipment* by Nicolas Brasch, Smart Apple Media, 2011.

*Secrets of Sport: The Technology That Makes Champions* by James de Winter, Capstone Press, 2009.

*Sports Technology* by Ron Fridell, Lerner Publications, 2009.

## Web sites

The official Formula 1 website:
*http://www.formula1.com*

A website with information about soccer balls:
*http://www.soccerballworld.com/*

The Hawk-Eye website:
*http://www.hawkeyeinnovations.co.uk/ Flasharea/Hawkeye.htm*

Describes an experiment to show the relationship of a golf club loft angle and the distance the ball travels:
*http://www.sciencebuddies.org/science-fair-projects/project_ideas/Sports_p013.shtml*

CBS Sports looks at the advances in golf ball technology over the years:
*http://www.cbssports.com/golf/story/10946668*

## Places to visit

Sole Technology Institute Lab Headquarters
Lake Forest, California
The world's first lab for studying the biomechanics of action sports, such as skateboarding and snowboarding.
*http://stilab.com*

Mechanical Engineering Department
Sports Biomechanics Lab
University of California, Davis
Davis, California
Learn about research on the motion of athletes and their equipment.
*http://biosport.ucdavis.edu*

# index